TECHNOLOGY 2000

TECHNOLOGY 2000

Facts On File Publications

New York, New York ● Bicester, England

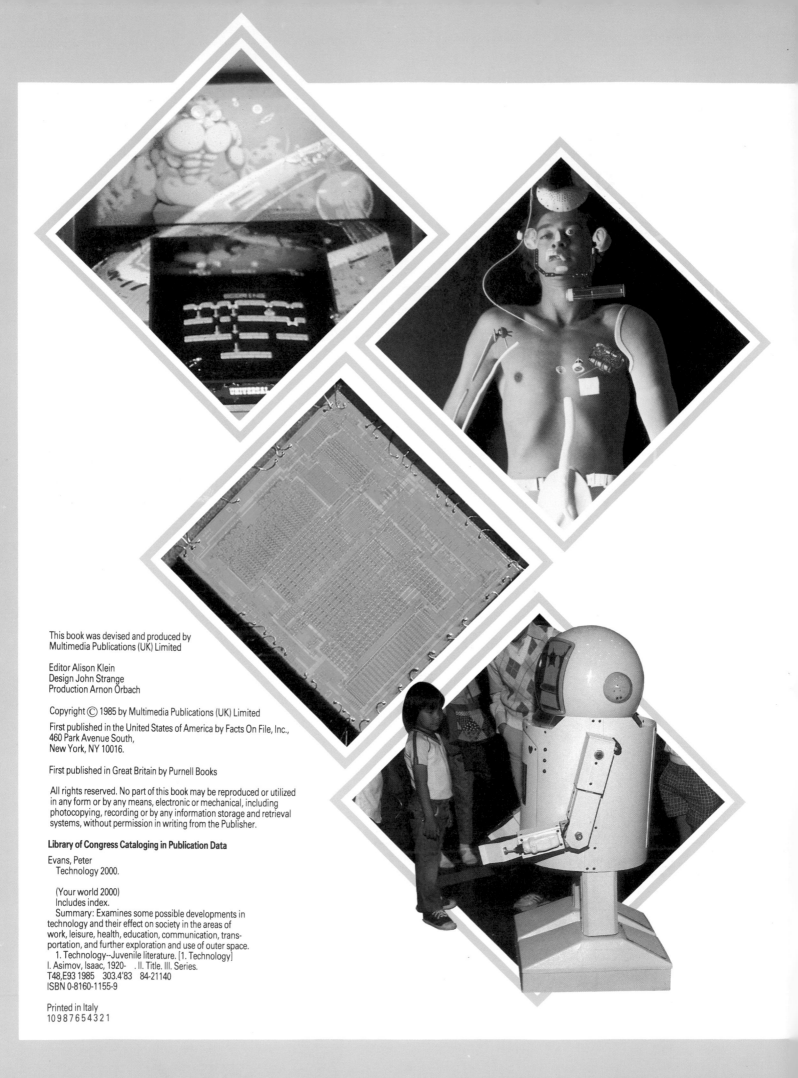

This book was devised and produced by
Multimedia Publications (UK) Limited

Editor Alison Klein
Design John Strange
Production Arnon Orbach

First published in the United States of America by Facts On File, Inc.,
460 Park Avenue South,
New York, NY 10016.

First published in Great Britain by Purnell Books

Library of Congress Cataloging in Publication Data

Evans, Peter
  Technology 2000.

  (Your world 2000)
  Includes index.
  Summary: Examines some possible developments in
technology and their effect on society in the areas of
work, leisure, health, education, communication, trans-
portation, and further exploration and use of outer space.
  1. Technology--Juvenile literature. [1. Technology]
I. Asimov, Isaac, 1920-   . II. Title. III. Series.
T48,E93 1985   303.4'83   84-21140
ISBN 0-8160-1155-9

Printed in Italy
10 9 8 7 6 5 4 3 2 1

CONTENTS

Foreword

As soon as any living creature begins to make tools, a form of technology comes into being. In this sense our primitive, small-brained ancestors created a technology at least two million years ago – they shaped little pebbles into cutting devices.

Through almost all of history, however, technology advanced only very slowly. When modern big-brained human beings came on the scene, 50 000 years ago, things began to move a little faster. By 1780, when an efficient steam engine was invented and what we call the Industrial Revolution began, technology went into high gear and new advances came more speedily.

They are still coming faster and faster – at breakneck speed. When I was young there were no television sets, no jet planes, no computers, no video games, no robots, no rockets, no push-button telephones, no hi-fi sets.

In the same way you will find that a great many things that you have never experienced will soon become part of your life, and make great changes in it. There may be great changes even by the year 2000, which is now less than two decades away. What will these changes be?

We don't have any crystal ball and, in any case, the future is always uncertain, but we can make reasonable assumptions. In this book we describe what the technological world of the year 2000 might be, if we proceed in a sensible way. But humanity must go peaceably about its task of developing and increasing its knowledge, and making use of it to increase the comfort and welfare of human beings.

As you will see, the world of 2000 may well be a very exciting place, and you will still be young enough to enjoy it when it comes.

Isaac Asimov

Isaac Asimov

Introduction

How will technology affect our lives in the year 2000? In this book we shall be looking at some of the ways in which technology will develop and how this will affect the way we live.

Computer terminals in homes could change the way people work and let them shop and bank without going out. Robot-controlled factories, teaching machines, electronic mail and information printouts could see a massive change in factories, schools and offices as we know them today. With the world's supplies of oil dwindling, we could be driving electric cars. Or perhaps individual cars will be a thing of the past, with fast, integrated transport systems operating.

The new technology could shorten the working week and give us more time to enjoy holograms, 3-D TV or computer games. With more leisure time we could well lead less stressful lives and enjoy better health. New drugs will give us better pain relief, with technology improving spare-part surgery as well as microsurgery. Lasers, video and satellites may be used in war as well as in peace to develop new missiles, aircraft and submarines and even to wage war from space. Or will we be able to use our knowledge of space to improve the quality of a peaceful life on Earth?

Peter Evans is a well-known science writer and broadcaster. As the regular presenter of BBC Radio 4's *Science Now*, he spends his time talking to scientists and technologists about current research and development. An Oxford graduate, he has written several books and many articles on scientific topics, mainly of a biomedical and technological nature.

CHAPTER 1
Information unlimited

We are in the middle of an information revolution that by the year 2000 will have completely changed the way we live. The key to these changes is the computer, basically just an adding machine, but an extraordinarily fast one. Today's computers are deaf and relatively stupid. But there are signs that a decade or two of research will make them see, hear and speak like humans. And they may be able to think for themselves, behaving like intelligent experts.

Computers will find their way into every human activity, communicating with each other in vast networks across the world. They will bring problems of privacy, and sometimes computers will oust humans from their jobs. But they also offer possibilities for education and recreation undreamed of until very recently. Computers offer us a challenge and a reward, provided we can make full use of their potential.

► Computers are now used in almost every aspect of our lives. In fact, we would feel quite lost without them. Their components vary considerably from the comparatively simple micro-chip of the domestic washing machine to the highly complex system of a flight control deck.

Research

Flight control

At home

Leisure

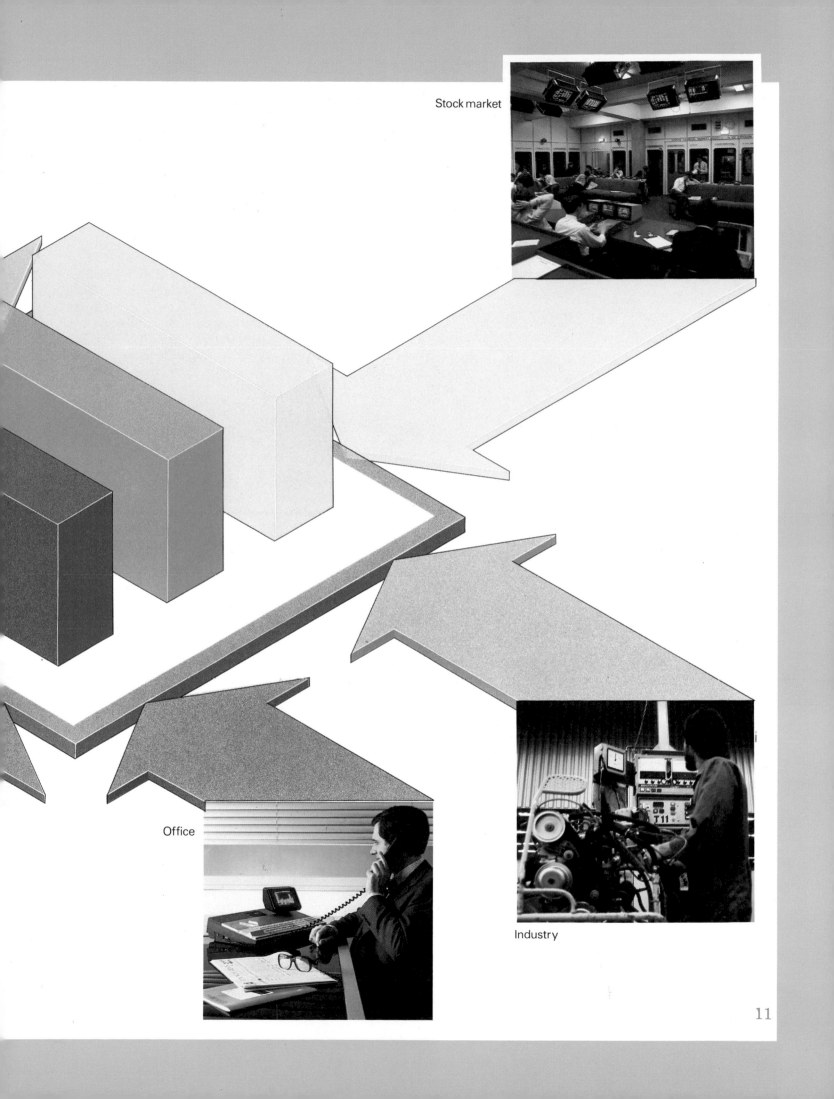

Stock market

Office

Industry

11

Information unlimited

▲ The abacus: addition, subtraction, multiplication and division can be performed rapidly by a skilled manipulator. Despite the arrival of the calculator, the abacus is still used today.

▼ Part of Babbage's analytical engine, designed to be programmed and able to store information in a memory. It was never built, because the technology of the time was too primitive.

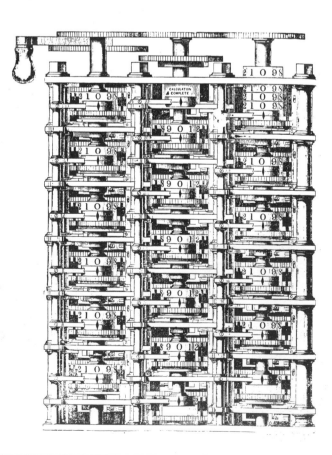

The rise of the computer

In the last thirty years or so, the computer – a machine for doing calculations – has gone from being an expensive rarity to a widely available cheap appliance. The earliest machines were enormously heavy and relatively slow-working devices, but today's computers are lightweight, compact and unbelievably fast.

No one could have foreseen just how useful and popular they were destined to become. In America, for example, the number of computers rose from 250 in 1955 to 20,000 in 1965. Ten years later there were four times that number in the US and roughly the same number in the rest of the world.

But these were all large 'mainframe' machines or 'minicomputers'. The real explosion came with the microprocessor of which 20 million were sold around the world in 1979. The present annual figure is thought to be reaching 100 million.

From abacus to micro

Machines to help us count and work out problems have been devised for centuries. Perhaps the earliest was the abacus – beads strung on wires which represent tens, hundreds and thousands in columns. A more modern idea was the slide rule.

But both these are manually operated calculating devices, whereas the long-standing dream had been of an automatic system. But it was not until 1832 that an eccentric Englishman called Charles Babbage produced a design for an 'analytical engine', very like today's computer.

Computers as such did not emerge until this century with a mechanical device called the Automatic Sequence Controlled Calculator (ASCC) in 1944, then the first electronic machine—the famous ENIAC (Electronic Numerical Integrator and Calculator) in 1946.

Pascal's calculating device, built in 1642, is a mechanical form of the abacus.

ENIAC was a gigantic affair, weighing 30 tonnes and containing 18,000 valves. Although these valves made for more rapid calculations, it was not until the advent of the transistor that really high speeds could be achieved.

Transistors, made from small pieces of silicon, were first used in computers in the early 1960s when they proved far more reliable and decidedly less bulky than the earlier valves. Computers were also becoming far cheaper to manufacture, a trend which continued when

An integrated circuit used in microprocessors such as the Commodore Pet and Acorn computers.

integrated circuits were introduced.

These are entire circuits of electronic components – the same as thousands of transistors – etched on to a single tiny chip of silicon. In other words, complete circuits can be made without having to solder together individual components. The most recent advances in chip technology are such devices as the 'transputer' developed by the British firm INMOS, which is really a whole computer on a single chip, made up of a total of 250,000 components.

Speed, size and cost

As processing speeds have gone up, so cost, weight and size have gone down. The early valve machines such as ENIAC cost around $1.4 million (£1 million) and filled a large room. The equivalent computing power of today can be contained on a single chip 5 mm square and 0.1 mm thick and bought in a micro costing a mere $70 (£50). And it can be slipped into a shoulder bag.

Compared to valve computers, transistor models were three times as fast, but a calculation that would take half a minute could today be completed in a tenth of a second. Researchers now are trying to make even

faster switching devices for computers including the very high speed Josephson junctions, which would enable many millions of calculations to be carried out per second.

Important strides will also be taken in increasing the computer's memory capacity. Storage is usually on some magnetic device such as a disk, tape or drum and is measured in kilobytes or K, each byte being 1024 separate items of information. One magnetic disk alone can hold the equivalent of 100,000 pages of printed information.

But to build computers to perform tasks such as holding a conversation or recognizing visual images, will take years of research into methods of holding information in store and retrieving it quickly for processing.

▼
A valve, similar to those used in the early electronic computers. Many thousands of such valves were used in these machines, which often needed a large room to accommodate them.

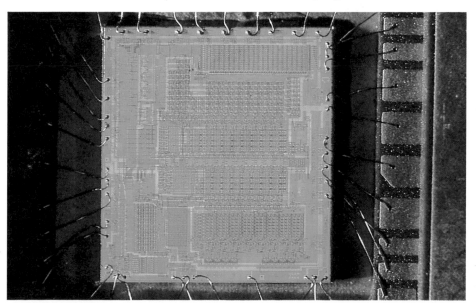

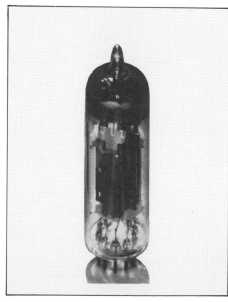

Information unlimited

Anatomy of the computer

Whatever their size, whether portable micro, desktop mini or roomsize mainframe, all computers have certain basic elements in common. They all consist of a machine or collection of machines for processing information. These machines are called the hardware. And they all need to be given programmed instructions, called the software.

If we follow the stages a computer goes through in handling a particular problem, we can see how the software and hardware are combined.

Imagine that the computer sits in the accounts office of a company where it prints out the salary slip for each employee every month. How does it do this?

Stage one: data input

The computer must be given data to work on and programs to instruct it on how to do this. This is done by a typewriter-style keyboard. The operator might enter the number of hours worked by everyone in the office, noting absences and overtime. Then program information is input, such as the hourly rate, allowances, tax codes and the insurance levels that are relevant.

Stage two: storage

The computer will hold on to the input in its memory, until instructed by the operator via the keyboard to process it.

Stage three: processing

When suitably instructed, the computer's central processor – the silicon heart of the machine – will input the data on hours worked and process this according to the program on rates of pay and tax codings. When the calculations are complete, these are sent back to the memory for storage.

Stage four: output

When required, the processed information is called for as output. It could be checked first on the visual display unit/terminal, VDU/VDT for short (which is like a TV screen) or it could be printed out immediately as a payslip with all the relevant information on it. An instruction could be sent by the computer to the bank to credit each employee's account with their wages.

Input keyboards, VDU/VDTs, output printers and other machines around the central processor are called peripherals.

Programs and languages

Because a computer, with all its calculating speed, can do nothing without a program to instruct it, programming must be done carefully to make sure that the machine proceeds logically and methodically.

Programs have to be written in special computer languages such as BASIC or FORTRAN, the choice of the language depending on the nature of the task. These languages use a binary code to represent numbers and letters. This makes them quite different from everyday speech, though there is a lot of research going on into 'high-level' languages, which are more like the way we speak.

Towards the fifth generation

In their short but spectacular history, electronic computers have gone through three 'generations': first, the vacuum machine, second the transistorized devices, and third, the integrated-circuit computers based on the silicon chip. People predict that two more generations are to come, the fourth being the era of even more densely packed microchips—very large scale integration or VLSI— and then the ultimate ambition, the so-called fifth-generation computers. These will come close to behaving very like human beings. Their artificial intelligence will enable them not only to see, hear and speak like us but they will also be able to reason and make deductions. These computers will only become a reality when we understand much more about the human brain and how it handles information.

▼ All data fed into a computer must first be converted into electrical pulses, either ON or OFF. Each ON or OFF, written as 1 or 0, is called a bit. A string of eight bits is enough to represent any single item such as a letter of the alphabet. The letter A is fed into the computer as one OFF followed by one ON followed by five OFFs followed by one ON. In binary code this looks like 01000001. The number one is represented as 0001. This way of representing the information may seem strange, but it uses the minimum number of bits.

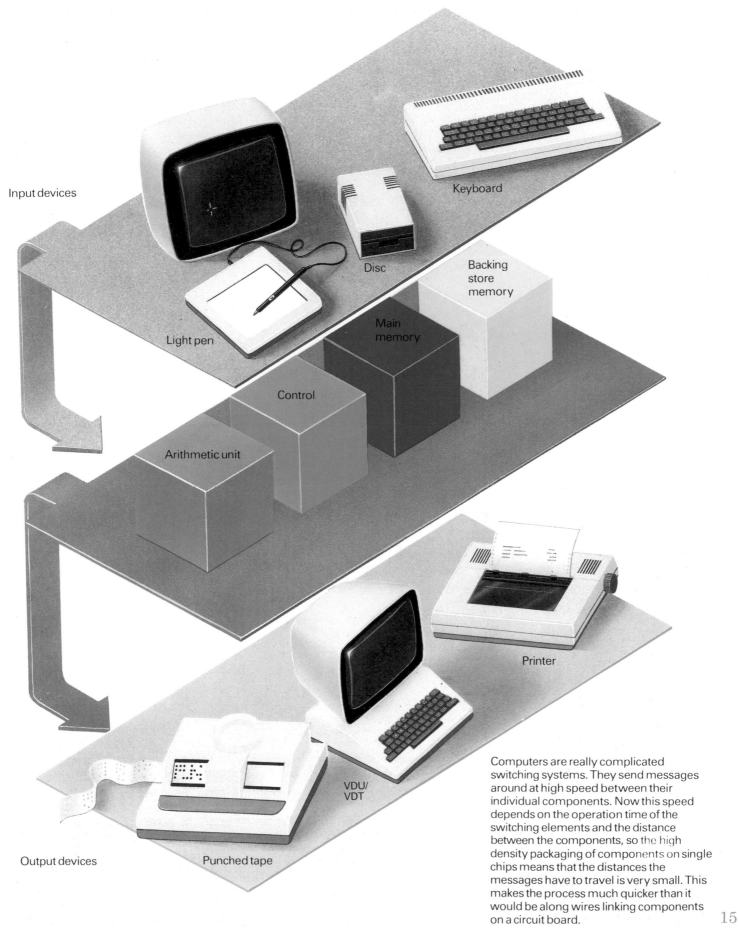

Input devices

Keyboard

Disc

Backing
store
memory

Light pen

Main
memory

Control

Arithmetic unit

Printer

VDU/
VDT

Output devices

Punched tape

Computers are really complicated switching systems. They send messages around at high speed between their individual components. Now this speed depends on the operation time of the switching elements and the distance between the components, so the high density packaging of components on single chips means that the distances the messages have to travel is very small. This makes the process much quicker than it would be along wires linking components on a circuit board.

15

Information unlimited

The computerized society

By the year 2000, computers will be so widespread and commonplace that we will have the same attitude towards them as we now have towards electricity or running water. We shall wonder then how we could ever do without them.

Many applications of computers that are starting to emerge now will be familiar by 2000.

Helping the police

Police can use the computer to match fingerprints at the scene of a crime with a set – if they have one – on their files. These files contain literally thousands of prints, which the computer will be able to sort through rapidly.

Another use for the computer in detective work is in providing eye witnesses with faces, to see if they can provide a picture of the person who committed the crime. Special artificially generated photographs are shown on a television screen. Then each feature is changed in turn by the computer until the overall face is close to what the witness saw. The police can then issue their computerized 'identikit' picture.

Personal records

As we move through life, from the moment we are born, we generate an enormous number of records: birth certificates, medical files, school records, passports and driving licences. We open credit accounts, pay computerized fuel bills, pay mortgages and rates, join clubs and societies and buy insurance.

In the future, all this data will be computerized for easy access. But this will raise the massive problem of security and privacy of information. Laws will have to be passed to ensure that certain records are not consulted without the individual being asked first.

◄ A blurred photo of a car number plate. After the photo has been processed by the computer, the number plate LUE 991L can just be made out.

► Records, records, records! Just some of your personal data which has found its way on to a computer.

▼ Crime fighting in America using a computer terminal in a police squad car.

Better banking

Banks will install more and more computerized cash points which will dispense money (provided of course that you have got some in your account!) either by reading your personal code entered by push buttons and identifying you by credit card or by recognizing your voice as you speak into a microphone. The same machine could also provide information on other banking affairs, leaving the bank staff free to carry out those jobs that require the personal touch.

Electronic mailing

Postmen will have lighter sacks to carry as mailing becomes a computer-based activity. A private individual or a secretary in an office will type out the letter on a word processor which will do two things: store the letter in the memory to serve as a 'file' copy; and send a 'top' copy to the recipient's word processor.

Computerized publishing

Not all information in the data-rich society of the year 2000 will be in the form of coded computer messages or VDU displays. There will still be a demand, probably even greater than today, for the printed word: newspapers, magazines and books.

What will change, though, is the method of producing the printed matter. The most laborious and expensive aspect of producing books or newspapers is typesetting the material. In future, word processors and computers will allow publishers to print the writer's final approved copy direct. Books such as this one will be put initially on to a word processor where a visual display or 'hard copy' will be provided for each page in turn. When each page has been approved by the author or editor it will be stored on magnetic disk, which can then be fed into a computerized photosetting machine. This machine will print out a perfect copy of the text. Newspapers will also be printed in this way.

And here is today's weather

Vast amounts of information on temperature, air pressure, wind speed, rainfall and so on will be gathered by ground weather stations, ships at sea, rockets, weather balloons, coastal stations and satellites. Today networks of computers already exist in England which can handle over 400 million calculations a second and provide analyses of weather information for TV weather forecasts, pilots and weather bureaus. In the future, more powerful computers will be able to give us even more accurate predictions of the weather.

CHAPTER 2
Transport and communication

To be able to go anywhere we want – near or far – in comfort, speed and safety is an ambition we all share. Tomorrow's transportation systems, using technologies that are now beginning to emerge, can make this possible. And if you do not want to go out in the world, you can bring it into your own home by a vast range of communications links – to inform, instruct and entertain. All at the touch of a button.

► Is this what is in store for us in the year 2000?

Transport and communication

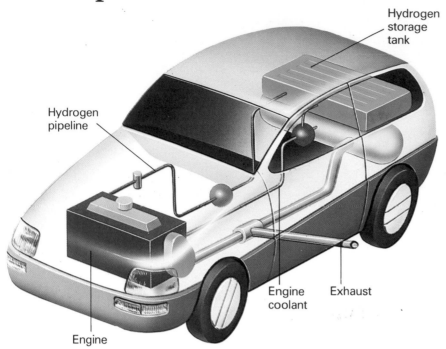

Hydrogen storage tank

Hydrogen pipeline

Engine

Engine coolant

Exhaust

◄ One answer to the problem of pollution may be the hydrogen-powered car. Liquid hydrogen, which has to be kept at a temperature below −253° C, is pumped to a specially designed engine. A microprocessor can be used to control the timing of the spark in the ignition system.

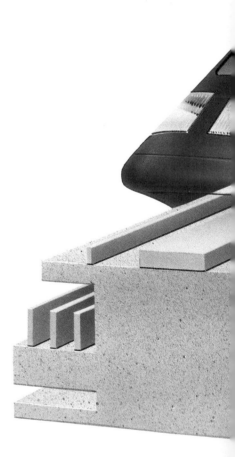

High-speed trains

Today we think it is a luxury to travel in quiet comfort in an inter-city train coasting along at 125 miles per hour (200 kilometres per hour). By the end of this century, though, this will seem almost cumbersome as trains reach speeds of 200 up to as much as 300 miles per hour (300 to 500 kilometres per hour).

Building trains to go fast is in itself well within the capabilities of today's engineers. As long ago as 1955, there were French locomotives that could top 200 mph (300 kph). But the important thing is to make these ultra high speed trips comfortable, especially around curves and bends, otherwise passengers will be thrown from side to side. There are several solutions to this problem. The first will be to modify existing tracks to iron out curves and make smoother junctions in much the same way as the French and the Japanese have already done with their sleek, aero-dynamically designed expresses.

The second approach to the speed-comfort problem is that already being developed in Italy, Spain and Britain – especially the last with its Advanced Passenger Train. As rebuilding tracks is very expensive, the British have instead designed the APT to take existing curves with greater ease than present trains. Tests show that using a tilting chassis allows bends to be taken 30 or 40 per cent faster, which means that instead of having to slow down to around 90 miles per hour (140 kilometres per hour), the train of the future can hold speeds in the region of 125 miles per hour (200 kilometres per hour) as it takes a typical curve.

Alongside these futuristic variations on today's trains, the year 2000 will also bring the third solution to the problem of speed and comfort: the monorail or the train without wheels. The train does not in fact come into contact with the track at all but floats above it, levitated by powerful magnetic fields generated by electricity. The train floats just a fraction of an inch (a few millimetres) above the track, which is enough to eliminate the friction that slows down more conventional rolling stock.

Today, prototypes exist for 'maglev' trains and you can see them in limited use at airports and fairgrounds. But for the principle to be commercially useable for long-haul passenger services, the technology will have to be developed to make the magnetic fields very powerful indeed. The answer will be found in the use of what are called 'supercooled' magnets. These generate magnetic fields by being cooled to such low temperatures that virtually all the current running through the coils is used. None is lost as heat because the coils have, with cooling, become superconductors. The monorail of the future with its superconducting magnets will be quiet and economical and reach speeds approaching 250 miles per hour (400 kilometres per hour).

Electric cars

The limitations and disadvantages of the internal combustion engine – noise, pollution, inefficiency, expense – have long been realized. So, too, have the attractions of the electric vehicle which by the end of the century will have become an everyday sight on our roads.

The small, battery powered vehicles, seen frequently in airports are limited in speed and range. Their batteries are relatively bulky and need to be recharged very frequently – usually after 50-65 miles (80-100 kilometres) of driving, sometimes less. What is needed is an alternative to the traditional lead/ acid battery, but even if this is not developed, it may be possible to combine the conventional internal combustion engine with an electric motor producing the so-called 'hybrid' car.

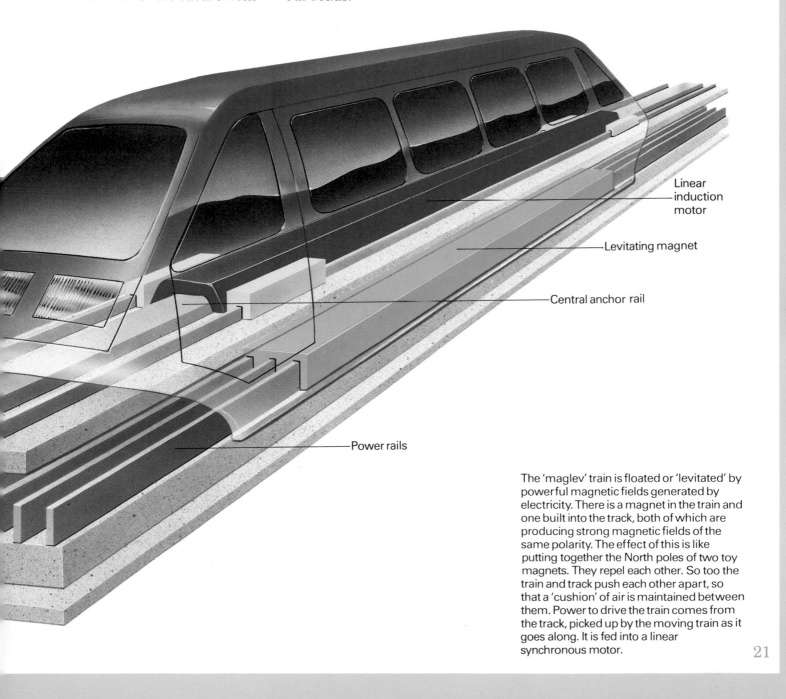

Linear induction motor

Levitating magnet

Central anchor rail

Power rails

The 'maglev' train is floated or 'levitated' by powerful magnetic fields generated by electricity. There is a magnet in the train and one built into the track, both of which are producing strong magnetic fields of the same polarity. The effect of this is like putting together the North poles of two toy magnets. They repel each other. So too the train and track push each other apart, so that a 'cushion' of air is maintained between them. Power to drive the train comes from the track, picked up by the moving train as it goes along. It is fed into a linear synchronous motor.

Transport and communication

Faster, safer aircraft

From the earliest days of man-made flying machines, there has been a constant attempt to build aircraft to fly further and faster. The passenger planes of the end of this century will undoubtedly continue this trend, but increasingly the emphasis will be on comfort and economy.

In general, the more seats on an aircraft, the cheaper the price of the ticket should be, so we can expect to see more super-airliners along the lines of Concorde or the Boeing 747 Jumbos. On the other hand, giant planes with many passengers create problems for airports, laying great stresses on runways and creating bottlenecks at check-ins and passport controls. So there will also be a demand for smaller aircraft, especially if these are less noisy as they take off and land over centres of high population.

Lighter than air

Alongside the more futuristic attempts to build faster, safer aircraft there will be a return to a flying technique that dates back to the early days of aviation: the airship. These fell from favour after several spectacular disasters such as the crashes of the R101 and the Hindenburg. But soon helium replaced inflammable hydrogen as the gas in the envelope and newer, tougher materials for the covering were developed. It became obvious that, provided you were not in a hurry, here was a cheap, quiet method of transportation, needing no runway to get it airborne.

Cargoes, too, may be carried eventually, at speeds around 150 miles per hour (250 kilometres per hour), thus making the airship in some cases a practical alternative to other forms of freighter with the added merit of being able to travel great distances without the need for refuelling.

Air control

During peak times, aircraft arriving at a busy airport are directed by traffic control to a stacking position – that is a particular level, separated from those above and below by 1000 feet, from which the descent is made to a lower level ready for landing, like a giant corkscrew flightpath. More and more, computers will be required to ensure that the flight controller receives critical information such as the plane height, position, speed, runway indication, range and so on, so that he can act with pinpoint accuracy. At present this information is on visual display, but eventually the computer may be able, by reproducing human speech, to 'say' it to the controller on demand if this extra facility is required. The pilot, too, may benefit from on-board talking computers relaying vital data about the flight and what manoeuvres should be carried out to keep it to schedule.

Controlling transport systems

Every day millions of people travel short distances in towns and cities by bus, car and subway. Increasingly the control of this busy traffic will be put into the care of computers, not only for regulating flow but for issuing tickets, opening doors, even driving the vehicles themselves.

San Francisco, California, already has a successful computer-controlled underground system called the Bay Area Rapid Transport system –BART. Its trains are fast, with speeds up to 80 miles per hour (125 kilometres per hour).

Road traffic, too, will be speeded up by more computer control. Already traffic-light systems are in operation to keep flow rate to a maximum. In the future, there could be automatic, unmanned Minitram cars holding about 20 people and shuttling around cities at speeds up to 45 miles per hour (70 kilometres per hour). These would run at predetermined intervals of say 5 minutes and as they stop and start would send signals back to the vehicles following to ensure that they ran in harmony.

As more aircraft ferry more people and goods into and out of airports, the problems of air-traffic control will multiply. We can expect to see improvements in radar, telecommunications and computer equipment to help the highly trained controllers to make take-offs and landings as safe and accurate as possible.

We can also expect to see dramatic developments in the transportation of freight by air. There are already designs on the drawing board for colossal freighters with 56 wheels and 12 engines. These giants of the air would carry payloads of oil, liquid natural gas or even water for barren regions.

Transport and communication

Telephones galore

Today you can pick up your telephone and enjoy a wide range of facilities: direct dialling and many information services on such things as the time, weather and travel. By the end of this century the telephone will be the core of an even more sophisticated communications system.

For one thing, it is expected that any domestic or business subscriber will be able to dial any other subscriber in any city, town, village or rural district in any country in the world. Then each telephone will have built-in features to make for easier call-handling such as: a memory for frequently called numbers, a forwarding facility to enable calls to be routed to you automatically if you are out, and a number display of the incoming caller so that you can set your phone to give off an engaged tone if you do not wish to take the call.

With the ever growing use of the telephone for business and domestic purposes it will become necessary for more automation to be developed in the handling of the enormously complicated network of calls.

Buried metal-wire telephone lines will also give way to the use of bundles of optical fibres – hair-like threads of special glass – along which far more channels of information can be sent than by the old metal-wire system. The information will be sent in digital form, transmitted by lasers – coherent beams of intense light.

By the end of this century, too, aided by fibre optic technology, the engineers will have produced a workable viewphone so that two callers in a conversation can watch each other's faces as they speak. Already there are prototypes for such machines.

Teledata: information on the small screen

Imagine being able to call up any item of information you require on any subject in the world: the birthdays of American Presidents, or the distance of the planets in the solar system or even just the day's news headlines. You would then see this instantly on your living room television set – all at the touch of a few buttons. This is the prospect offered by various systems for TV data display that have already come into service in a limited way, but which could easily be extended to every home.

Satellites

The vast increase in communications traffic by telephone and television will see a growing number of specially designed satellites launched to handle the flow.

Obviously there is a risk of collision. Thousands of orbiting satellites, some of which are no longer being used could produce a space traffic jam. So in the future the reusable Space Shuttle may, as well as launching satellites, have to act as a special kind of refuse collector, scooping up unwanted orbiting garbage.

Although satellites are expensive to launch, they will be increasingly relied on to handle communications links. Undersea telephone cables are not only costly but incapable of meeting the expected demands of the next century. Television broadcasts of important events such as the Olympic Games or Royal Weddings are wanted 'live' on the other side of the world. Satellites provide the ideal method of relaying these without being subject to atmospheric variations that could ruin the transmission. And as for telephones, an orbiting satellite can at present handle as many as 13 000 calls at any one time, making it a vital adjunct to existing land line systems.

Radio dishes to collect satellite information will be as familiar as TV aerials are today. The big picture shows a communications satellite seen through the window just after it was launched by the Space Shuttle.

▼ Information in our homes at the touch of a button is already available through commercial systems. By using these links we might soon be able to do our shopping from a cosy armchair, buying anything from a bar of soap to a new house.

▼ Computer games, too, can be at the end of a telephone line, using another form of data by TV. Using a keypad, you will be able to call up a central computer and get the information you want in the form of text or a diagram.

▼ It may soon be quite common for meetings between various branches of a firm in different towns to take place on the telephone. The speakers will be able to see each other's faces, saving the time and expense of travelling.

25

CHAPTER 3
Health and medicine

A lifetime of good health for everyone is the target of doctors and medical researchers. The aim is not to try to live forever, but to give us all the means to enjoy as long and healthy a life as possible. New drugs will be introduced to help stave off the degenerative diseases, such as arthritis and cancer, and the mental disorders of old age. New technological tools will help doctors to diagnose illnesses earlier and more accurately, while the sick will be cared for with greater efficiency and given more chance of a full recovery. That is the hope and the expectation.

➤

Today grandparents and grandchildren living in the west have the opportunity to do lots of things together. As technology improves, it is hoped that many old people can look forward to such fulfilling and pleasurable lives.

Health and medicine

Towards the bionic man and woman
The real Six Million Dollar Man will not be springing into action by the end of this century. But here is a possible workshop manual to guide you round the repairs and replacements that may be available to the body machine in the year 2000.

The brain
The human brain is unimaginably complicated. There is no chance of replacing this with, say, a silicon-chip based computer. But partial brain grafts – taking brain cells from one area of a donor brain and injecting them into a host – are a possibility, perhaps to remedy brain damage after an accident or to stop the onset of mental deterioration in old age.

The eyes
Nowadays grafts of the cornea – the clear window at the front of the eye – are commonplace. But the hope is that the next twenty years or so will see eye surgery extended to restore sight to the blind, or at least some of them. Tiny, miniaturized TV cameras would be linked to the brain where visual information is processed. This would allow the blind to 'see' without using their natural eyes.

The ears
Present hearing aids only work if the correct nerves in the ears are still functioning. In the future it may become possible to help the totally deaf by electronic means, implanting minute electrodes in the inner ear which transmit sound impulses to the brain.

Speech
Although bionically engineered speech is likely to be more difficult to achieve than hearing or vision, there are signs that prototypes could be available before the year 2000. Here the idea is to implant electrodes in throat muscles that would pick up the electrical impulses that come just before speaking. These impulses are relayed to a tiny processor which converts the signals into human speech – a miniature talking computer.

The heart
Ever since Dr Christiaan Barnard and his team proved, in 1967, that heart transplants using hearts taken from human (dead) donors were possible, these operations have become increasingly successful. A major problem, though, is rejection – the receiver's body tries to oust the 'foreign' tissue. In the year 2000 drugs for dealing with this are to be expected. Another problem lies in getting a supply of appropriate hearts to the right place at the right time and this is where artificial organs will come into their own. Steel and Teflon hearts are already being used successfully.

Kidneys
At present a person with severe kidney disease can either rely on a dialysis machine or undergo a transplant operation, provided a suitable donor is available. Again the trend is towards effective artificial organs. Already a small wearable synthetic kidney is being tested.

Liver and pancreas
What are today just plans for an implantable artificial liver may also be realized by the end of the century. But even if they are not, efficient and safe back-up systems to the natural liver should be available.

Diabetes sufferers whose pancreas fails to produce the hormone insulin will also be helped. There are already models for an insulin infusion pump worn on the belt to deliver the hormone when required.

Limbs
Artificial limbs have developed greatly since the first relatively crude aids. Now bioengineers are drawing on advanced technologies in materials science, electronics, mechanical engineering and so on to build limbs that provide increasingly natural movement. In the future, though, the prospect is for artificial limbs that derive their movement from existing muscles, or, even more exciting, from the nerve impulses that would drive a normal hand, arm or leg.

►Apart from the many parts of the body that can be replaced today, such as the heart, kidneys, liver and various joints, in the year 2000 we may see replacement of small areas of the brain, tiny cameras to replace the eye, electronics fitted inside the ear to help the deaf and even synthetic artificial kidneys or hearts.

The man on the right shows the range of spare parts available today: plastic arteries, alloy and polyethylene joints, heart valves and pacemakers, artificial eyes and ears, rubber testes...

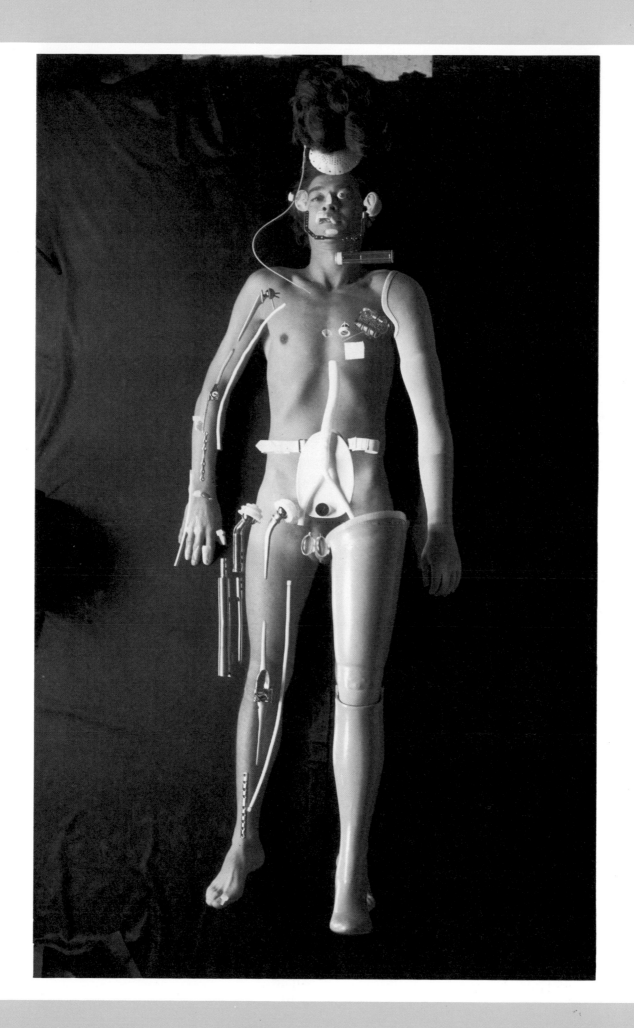

Health and medicine

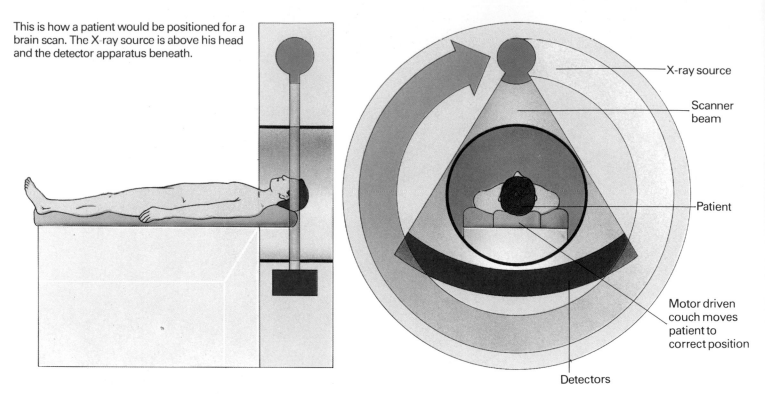

This is how a patient would be positioned for a brain scan. The X-ray source is above his head and the detector apparatus beneath.

X-ray source

Scanner beam

Patient

Motor driven couch moves patient to correct position

Detectors

Inside the operating room

The modern operating room is a remarkable man-made environment, very carefully designed to allow large teams of surgeons, anaesthetists and nurses to carry out their various tasks quickly and efficiently. At the same time they have to ensure that the patient runs the least possible risk of infection to the wound.

We can expect to see a number of developments in all aspects of the operating room's functioning: a better, cleaner air supply; better materials for making the caps and gowns; and even more sophisticated operating tables.

There will also be innovations in the surgeon's 'tools of the trade', some of which have already begun to emerge in a limited way. We can expect developments in computer surgery, where microprocessors control the action of tiny, rapid bursts of electricity used to sever tissue or cause blood to coagulate, in laser surgery and in cryo-surgery – in which the surgeon's tools are at low temperatures.

Computers on call

As elsewhere, the computer will figure largely in our medical lives: helping to diagnose illness, train medical students, save lives in emergency, give check-ups and ensure that doctors and nurses have all the information they need to give us the best possible treatment.

Much of a doctor's time – in hospital or in general practice – is taken up with getting a medical history from a patient to help in the diagnosis and treatment of the illness. In the future there will be more and more use of computers such as the Mickie machine. It can converse in an agreeable fashion with the patient, asking questions about the details of symptoms and how long they have been experienced.

Tomorrow's medical students may not need to use passive, unresponsive anatomical models while they are training. Instead they will have access to remarkable dolls, such as Sim One, which can react to certain drugs just like a human being, altering breathing rate and blood pressure – even changing the blink

▲ On the outside looking in. The body scanner allows doctors to make internal examinations without even touching the patient. Detectors take over a million readings to form the final picture.

rate and pupil size of its eyes. A computer print out produced by the same processor that controls Sim One will help students assess their performance.

Doctors will also be able increasingly to get a full medical history of anyone in their care by tapping out the appropriate access codes on a VDU/VDT keyboard. Up will flash the data the doctors need, saving time, and perhaps even lives. With the networking of computers and easy interchange of information over telephone lines, this will be possible if the doctor is in his office one side of the country and his patient's history in a hospital file on the other, hundreds or even thousands, of miles away.

Conquest of pain

Controlling the distress of pain has always been a major concern of doctors. In the future, the hope is that new generations of powerful pain-killing drugs will be developed without unwanted

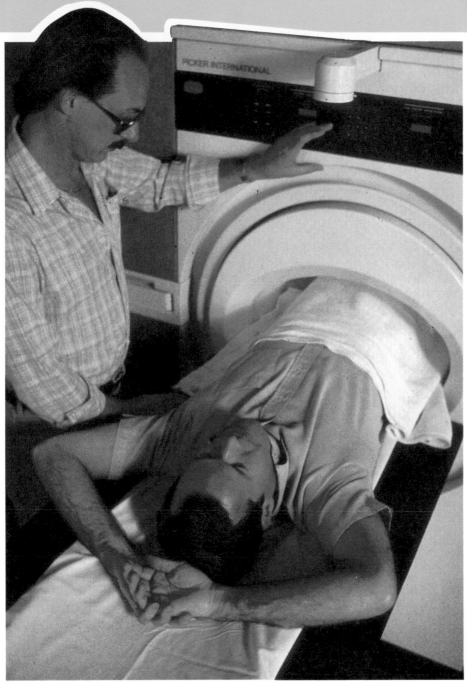

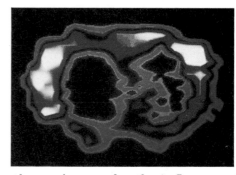

◄ Lining up a patient in an NMR scanner. The exact position is determined using a laser beam.

▼
An NMR picture of a diseased lung. Part of the lung space has been taken up by a growth.

side effects or the possibility of addiction. The discovery in the mid-1970s that the brain produces its own pain killers – the endorphins – may be just the clue researchers have been seeking to make this possible.

Scanners

High-technology scanners will become familiar items in tomorrow's hospitals and clinics, making present methods of examination, diagnosis and research look decidedly old-fashioned.

CAT scanners (computerized axial tomography) can take a series of X-rays from different angles through a thin section of the body and process these through a computer to provide startlingly revealing pictures of abnormalities such as brain tumours that, before, could only be discovered by surgery.

NMR (nuclear magnetic resonance) is another of the new scanning techniques, less developed as yet than the CAT method, which will grow in importance as a way of seeing inside the body, locating cancers, say, or lung defects. It works by picking up the minute signals produced by atoms as they align themselves under the influence of a strong magnetic field.

PET (positron emission tomography) is a bit like the CAT scanner, but its detectors respond to atomic particles called positrons, not X-rays. It is a valuable tool for investigating illnesses such as cancer, strokes and epilepsy.

BEAM (brain electrical activity mapping) scanners are also being developed to enable doctors to monitor what goes on in the brains of people with illnesses such as schizophrenia, tumours, epilepsy and even, perhaps, reading disorders.

Health and medicine

Unlimited drugs with biotechnology

The human body is like a factory, making many vital chemicals to enable it to survive and grow. It is now becoming possible to produce some of these biological substances in the laboratory using the revolutionary methods of recombinant DNA technology – better known as genetic engineering. As we move towards the end of this century, genetic methods of manufacturing medically useful drug chemicals will spread rapidly, providing new hope for people with crippling diseases. It might also lead to cures for disorders that up to now have been virtually untreatable.

To understand genetic manipulation you must first know something about the structure and behaviour of the living cell, of which the body has many thousands of millions. Each cell of every organ of the body is a microscopically small self-contained unit, with three principal components: the outer wall or membrane, the jelly-like cytoplasm, and the nucleus floating within it. Within this nucleus are the cell's genes, arranged within groups called chromosomes. Each gene–made of the molecule called DNA – has its own special role to play in regulating the behaviour of the cell, governing its growth and the various substances, such as proteins, that it manufactures. Thus genes within the nucleus of cells in the pancreas will be responsible for making quantities of the hormone insulin, which is needed to break down sugar in our food.

But insulin is also a useful drug, because it is used by diabetics to control their sugar balance – their own pancreas being unable to manufacture the hormone. So it would be extremely valuable to be able to produce this in great quantities. The same is true for other substances such as growth hormone, or interferon, the body's own anti-infection agent. And this is where genetic engineering comes in.

The gene responsible for insulin or growth hormone, for example, is first isolated. Then it is introduced into a special bacterium – a single-celled organism – where it will 'express' itself as it would in a human body cell, by making a minute quantity of the required substance. The bacteria are then grown, multiplying rapidly, each one of the many millions that result acting as a tiny hormone factory. In this way quantities of a natural version of a human drug are produced.

Insulin, growth hormone and interferon are the first useful substances to be manufactured in this way. Later it is expected that a vast range of drugs will follow, including those to regulate fertility, to stimulate failing memory, to control appetite, and to promote sleep and kill pain.

Reading genes for better health

Many diseases are genetic in origin. That is, they are caused by some abnormality or defect in the DNA within the cell nucleus. Already it has become clear that the genes can to some extent predict a person's medical future. Some genetic diseases can be forecast by looking at the

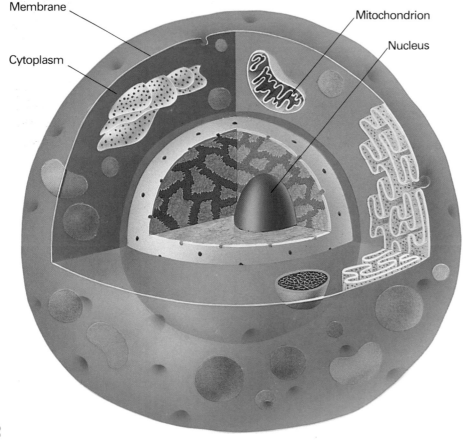

Membrane

Cytoplasm

Mitochondrion

Nucleus

◀ The structure of a cell, many thousands of millions of which go to make up our bodies. The cell is made up of an outer wall or membrane, the cytoplasm inside the cell and the nucleus, the command centre, floating in it. Genetic material is found in the nucleus. Some is also found in the mitochondria, the cell's power house.

▶ The baby's head seen through the ultrasound scanner. The size of the baby's head is measured from the picture and can be compared with a standard chart. In this way it is possible to monitor the baby's growth. Genetic monitoring in the future may allow doctors to perform operations on babies before they are born, to prevent certain defects.

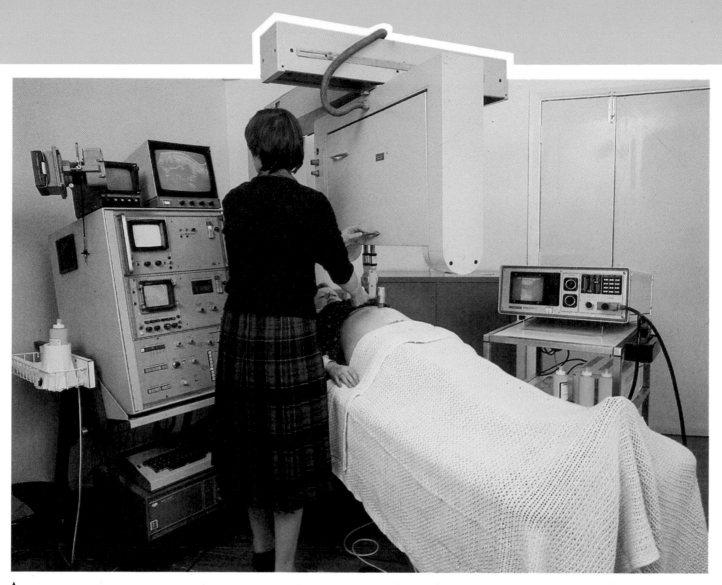

▲
The ultrasound scanner uses high-frequency sound waves to form an image. Here it is being used to monitor the progress of an unborn baby in the womb.

You can see the picture on the TV screen. In the future, it may be possible to examine the unborn baby's genes to tell if he or she will get certain diseases later in life.

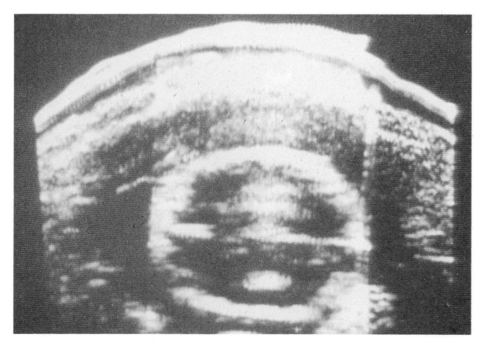

individual's genes.

It should become possible – perhaps by the end of this century – to examine the genes of the unborn baby, very early on in pregnancy and tell which, if any, diseases he or she is likely to succumb to.

Drugs on target

If you take a pill or have an injection, the medicine enters the bloodstream where it is circulated around the body, eventually ending up at the site where it is needed. This is a fairly crude way of targeting the drug – a bit like throwing a handful of darts at the board and expecting, or perhaps just hoping, one will hit the bullseye.

By the end of this century drugs will be targeted far more accurately, meaning that they will be more effective and that smaller doses will be required.

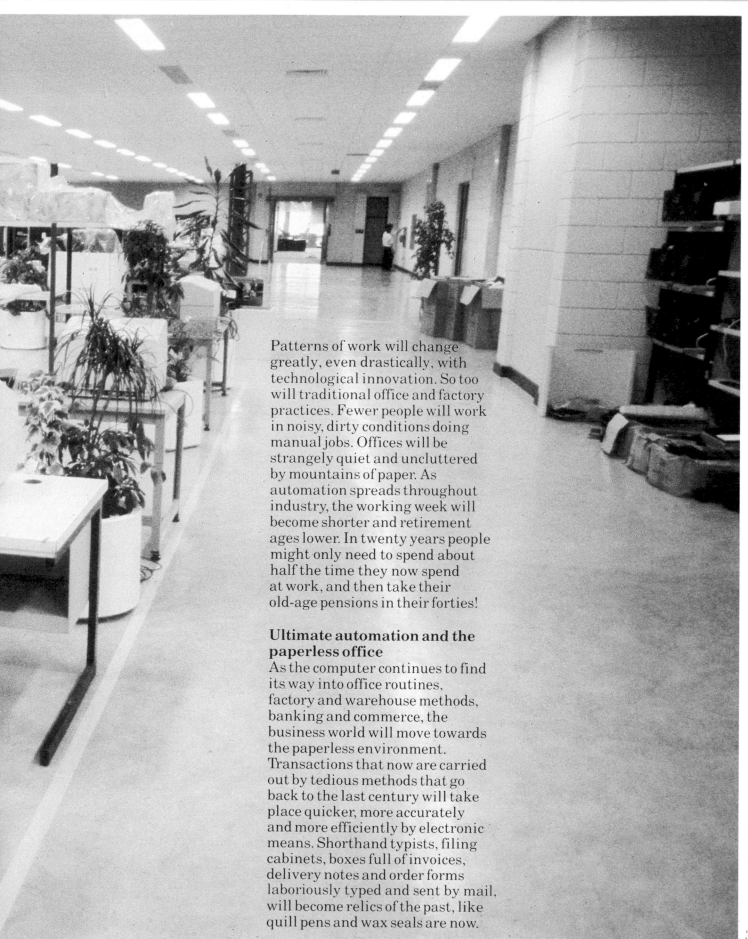

Patterns of work will change greatly, even drastically, with technological innovation. So too will traditional office and factory practices. Fewer people will work in noisy, dirty conditions doing manual jobs. Offices will be strangely quiet and uncluttered by mountains of paper. As automation spreads throughout industry, the working week will become shorter and retirement ages lower. In twenty years people might only need to spend about half the time they now spend at work, and then take their old-age pensions in their forties!

Ultimate automation and the paperless office

As the computer continues to find its way into office routines, factory and warehouse methods, banking and commerce, the business world will move towards the paperless environment. Transactions that now are carried out by tedious methods that go back to the last century will take place quicker, more accurately and more efficiently by electronic means. Shorthand typists, filing cabinets, boxes full of invoices, delivery notes and order forms laboriously typed and sent by mail, will become relics of the past, like quill pens and wax seals are now.

At work

◄ This robot device is for handling dangerous radioactive materials. It's better than risking the life of a human operator!

The traditional production line will virtually disappear as automated systems and robots take over repetitive jobs. Released from dirty, unrewarding tasks, which can also be dangerous, the human work force can be trained to higher levels for more rewarding posts, where human abilities are unrivalled.

A microcomputer makes it easy for the shopkeeper to check and update stock. Electronic recorders at the till record sales automatically and deduct each item from the list of stock, as well as issuing receipts.

◄Computerized warehousing. Heavy lifting of boxes on to vans and lorries will disappear from the list of human jobs as robots do this more efficiently. But the driver who coordinates hand and eye, uses judgement and experience, recognizes and responds to danger signals and hazards will certainly not be replaced by a machine for a very long time to come.

Cash by computer. Banks are already using cash machines for customer convenience. In future, cash will be available from cash machines 24 hours a day, provided of course you have enough money in your bank account!

Robots may not be the best of musicians, but they can be used for testing instruments as they are manufactured.

At work

Robots come of age

With surprising suddenness the robot – a machine that, in some respects, can act like a human – has become an everyday reality. This is especially true in industries such as automobile manufacturing where robot 'employees' are busily casting metal parts, welding joints, inspecting for faults and spraying with paint.

But the robot is a tool with a difference. Unlike a socket wrench or a sewing machine, it is able to take in information about its environment and interpret this in such a way as to guide its own decision making. Robots in short can observe, interpret and act, which is why these sophisticated, computerized tools, though stupid by human standards, can be said to come some way towards the artificial person.

In the year 2000 robots will be well established in our daily lives; in factories, offices, banks and supermarkets, in space and in our homes. The increase in their population will be phenomenal. In 1980 there were reckoned to be a total of 8000 robots in the entire world. The graph on page 39 shows one estimate of robot numbers in the year 2000.

A robotized society

If we are to use robots on a wide scale, and if technological advances towards the 'thinking' computer continue, it will become necessary to lay down guidelines governing where and how they are used.

Perhaps by the end of this century we may have to pass something like the famous Three Laws of Robotics put forward some years ago by Isaac Asimov. All robots would be built containing programs to ensure that these basic rules were not violated.

1 A robot may not injure a human being, or, through inaction, allow a human being to come to harm.
2 A robot must obey the orders given to it by human beings except where such orders would conflict with the First Law.
3 A robot must protect its own existence as long as such protection does not conflict with the First or Second Law.

Fact not fiction

Robots in the real world do not look a bit like some of the robots we have grown used to in films and stories. But they are still extraordinary devices and far cleverer than machines of earlier times.

The robots that now work in industry depend on a program which tells them how to perform a particular task in a particular way.

Among the simplest jobs is 'pick and place' where a robot may take a heavy component from one machine and move it to another for further processing. The main lifting arm is driven by pneumatic mechanisms for reaching outwards, and has at the end a gripper unit which can manipulate by swivelling and by opening and closing.

For more complex jobs robots are equipped with servomechanisms. These are designed to get the robot to 'respond' to situations and act accordingly. For example a robot may be built to insert a component in an oven and remove it when it reaches a certain temperature. Thus the robot has both to measure the temperature and automatically withdraw the component. In other words it has to react to 'feedback' from the environment.

Robots too need memories for more complex tasks. These are electronic in form, stored on tapes or disks.

The robot arm is first guided manually through the sequence that is wanted. As this is being done the robot memory records the sequence precisely. It can now repeat the procedure on its own. The teaching sequence then becomes the robot's computer program.

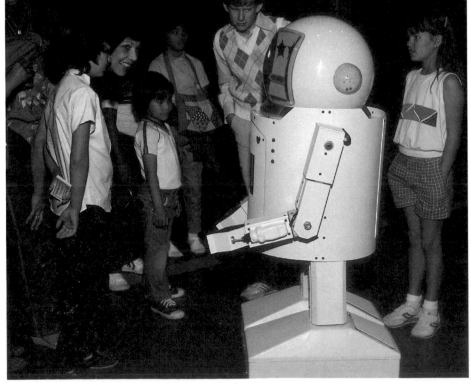

◀ A robot called 'Denby', not very clever but great fun to watch. Really useful robots do not look much like Denby.

We can expect tomorrow's robots to be more versatile and more 'intelligent' than those of today. They should be able, in a limited way, to recognize shapes and to 'feel' with very high sensitivity sensors. They will certainly end up understanding speech and talking back in something resembling human tones. They will eventually be able to move around more freely than today. They will not be rooted to one spot all the time. And if they go wrong they will be able to diagnose their own faults and tell human (or other robot) operators which parts need repair or replacement. Clever 'second generation' robots are already taking their first steps. The year 2000 will see them, in certain fields, becoming accomplished, mobile performers.

▲
Robots at every stage in the production line will manufacture the car of the future.

▶ C-3PO, the human-like character from *Star Wars* – a robot with millions of fans.

▼
In just a few years there has been a spectacular rise in the robot population. Here is a prediction for the year 2000.

| Brazil: 550,000 | Sweden: 650,000 | UK: 820,000 | E. Germany: 1,000,000 | Italy: 1,600,000 | France: 1,620,000 | W. Germany 3,600,000 | USSR: 5,600,000 | USA: 7,500,000 | Japan: 11,000,000 |

39

At work

Changing traditions

Technological advances and innovations will mean that some jobs that people do today will, over the next twenty years or so, vanish altogether. At the same time, other, completely new jobs will become available as more ultra-modern industries emerge.

Comings and goings

Many jobs will be out-of-date, including machine operators on lathes, grinders, paint sprayers, fitters and riveters, loaders and packers. In the office, typists, mailroom staff, art and design staff for pasting up pages, filing clerks and book and magazine librarians will no longer be needed. Elsewhere door-to-door salesmen, ticket-office clerks and collectors, bank clerks, and telephone operators will also be jobs that will disappear.

New kinds of jobs will include genetic engineers and enzyme experts, bio-engineers to produce artificial limbs and organs, and networking technologists to set up the complex computer systems in factories and offices and ensure that they operate properly.

Computers: from cartoons to car parts

Just as robots will take over the repetitive handling tasks in the factory, so too will computers be able to help in jobs such as design and illustration that have their share of laborious activities.

A good example of this is computerized animation. To make a cartoon film by traditional methods is a slow process. Each tiny movement of the cartoon characters has to be represented in a different drawing which is then photographed on one frame of film. When the film is run through the projector at the rate of 24 frames a second this gives the familiar illusion of movement. But it takes no fewer than 1440 drawings for every minute of finished film. A full-length cartoon movie therefore can require hundreds of thousands of individual drawings, involving thousands of hours of work for the artists. The computer can speed all this up with far less labour. It can be programmed to reproduce slight variations in an original drawing, called image manipulation and also to operate the rostrum camera that shoots the drawings in precisely the correct sequence. Not only does this save the artist from making the thousands of drawings individually, but it produces a much smoother flow of images.

Computer aided design (CAD) will also be used more and more in the industrial setting. Here the idea is to program the computer to

▼

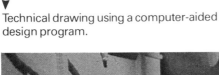
Technical drawing using a computer-aided design program.

► The farm of the future, powered by an array of wind turbines generating electricity. A computer-operated robot machinery is in the fields, crops are grown in specially controlled environments and the farmer is supervising it from a control tower.

provide three-dimensional displays of a given drawing from a variety of angles. So a design engineer may for example want to 'look at' a particular car component before it is made in metal, to study its features in detail. This he can do with the computer.

Down on the farm

High technology will invade agriculture. Tomorrow's farms will bear little resemblance to the traditional image of farming.

For one thing, genetic engineering will lead to specially devised crops that possess all kinds of useful qualities. There might, for example, be corn that automatically comes into bloom each year like a perennial flower, without the need for seasonal replanting. Another possibility is the development of special hybrids such as a potato and tomato that both grow on the same plant. Or – one of the most enticing ideas of all – a plant engineered to contain all the basic nutrients needed for a balanced diet. The perfect food crop.

Animal rearing too will be improved: pigs, cows and sheep specially bred to be immune to common diseases would be one important advance. So too would be livestock producing more meat per animal. Again genetic engineering will help in this respect.

Computers too will be out in the fields, with sensors to monitor such features as lack of soil moisture. In this way the farmer could detect and put right deficiencies without having to send out employees to see for themselves.

In short, with the aid of computers in all aspects of the farm's life, there will be a revolution on the land.

CHAPTER 5
Leisure and education

Technology will give us an unprecedented opportunity to enrich our lives through education and self-improvement, and to enjoy to the full our moments of leisure.

The vast wealth of knowledge in the world's libraries, museums, galleries and universities can be brought effortlessly into our homes. From our very first days at school we can enter a world of experience and stimulation, acquire skills and be helped when we are slow to learn.

And we can feed our imagination and fantasy as well with entertainments and amusements of the most vivid kind – from personal brainwave games to a guaranteed seat at the Centre Court at Wimbledon on Finals day.

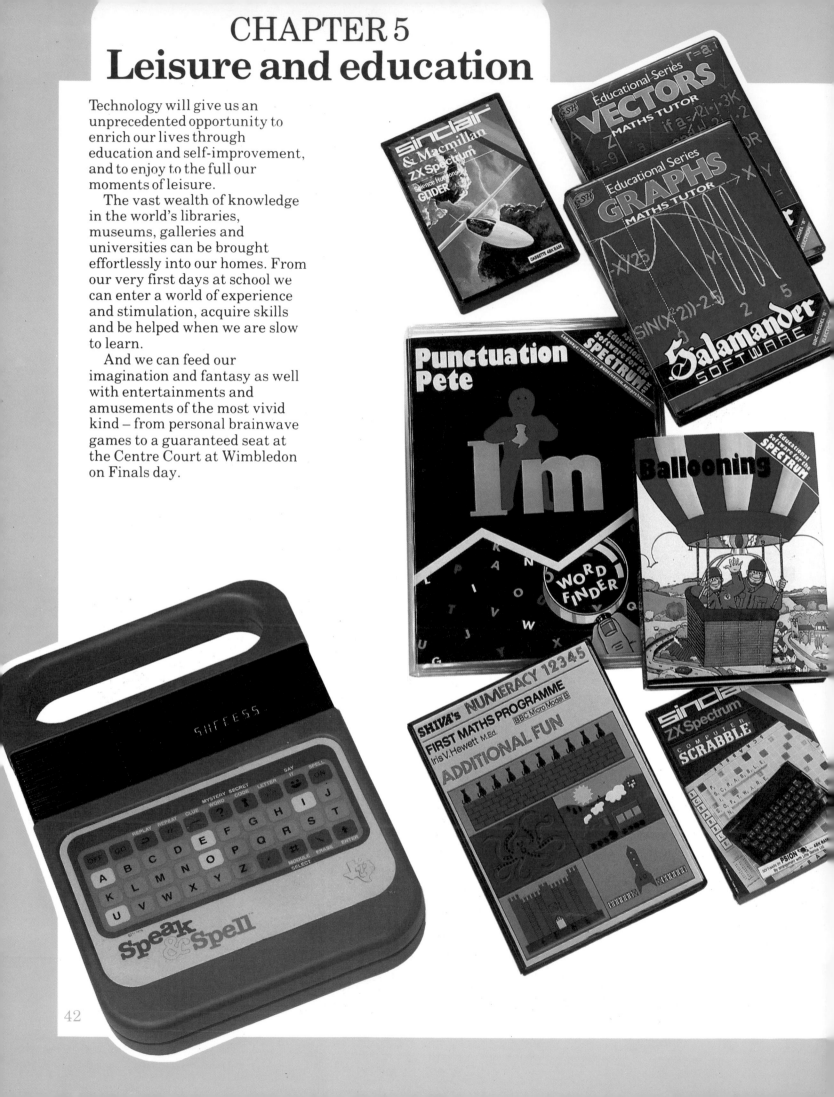

Leisure and education

Classrooms of the future

Nowhere will the spread of information technology – IT – benefit society more than in the field of education. Already we can see trends emerging that are sure to continue: the growing use of stored data, on film and tape, to bring lessons alive in the classroom; greater awareness of the value of computers in our daily lives; the need to know how and where to get information when required rather than having to carry it all around in our unreliable memories; the development of special teaching aids for those who are slow to learn or who have a handicap that impairs them.

▼

High-tech teachers. Tomorrow's good teachers will be the same as they are today, well informed and able to impart their information clearly and concisely. But they will also have a range of helpers in the shape of electronic aids. This means that in any classroom they will be able to run effective teaching groups simultaneously, spending their own time helping the slower learners, sorting out problems and generally overseeing progress. If a child is off sick from school, there will be no need for school lessons to be missed. Once well enough, he or she can link up with the teacher. Units of work will be transmitted down the telephone line and homework will be marked by the same channel.

► A French cartoon drawn in 1900 and depicting an imaginary classroom in the year 2000. How close to the truth do you think it will be?

▼
Electronics can be great fun. These primary school children are obviously enjoying making the electronic 'turtle' draw.

Leisure and education

Home box-office

Long gone are the days when, every night of the week, people would wait in line at the cinema to watch the latest movie. This may still happen when spectacular 'big' films are released that require a large theatre and screen for their effects to be appreciated. But for the most part, television has become the cinema of the modern world. The arrival of the video cassette recorder, or VCR, has meant that now you do not have to wait for a movie to be broadcast. You can hire it or buy it from a shop. The home box-office concept has emerged.

Over the next twenty years cheaper VCRs will mean that this item of hardware will become almost as common as the television set with which it is used. In 1981 it was estimated that in America there were just over 2 million VCRs in people's homes. By the year 2000 the expected total is ten times that amount.

It is also thought that the videodisc – records that play pictures and sounds – will be almost as numerous, especially those 'played' by fast scanning beams of laser light.

There will be more and more television piped directly into homes by cable. Subscribers will pay a certain sum of money to have programmes of particular interest to them fed by cables into their houses and apartments. So if, say, you were very fond of yoga, gardening or ballet you could choose to have programmes on these topics in abundance.

Satellites in geostationary orbits (their position in relation to the Earth never changes) will also be a source of more TV programmes. Direct broadcast satellites or DBS will relay dozens of channels into every home, the signals being picked up by small radio dishes on the roof.

Three-D entertainment

One of the most exciting ideas in the field of entertainment to emerge in recent years has been the hologram. This is a three-dimensional image, showing not only length and breadth but also depth, produced by the clever use of a laser beam.

In future we can hope to see two major developments in holography. Present technology only allows for three-dimensional holographic images with limited movement and colour.

However, the cinema of the

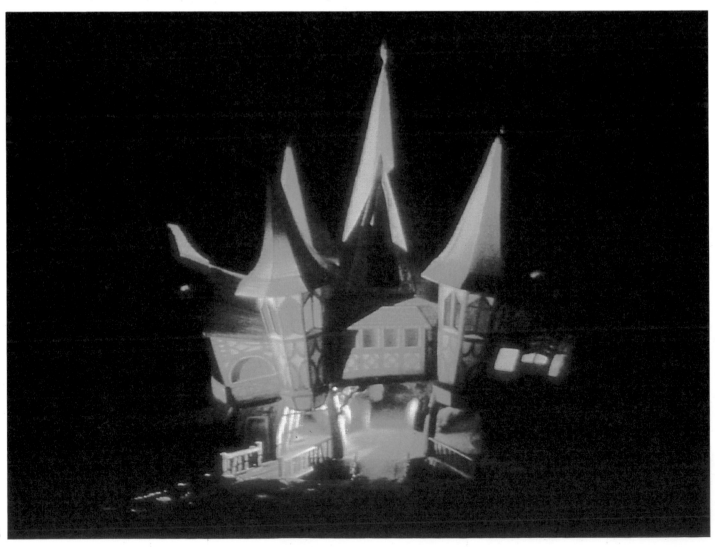

year 2000 could well be showing holographic movies, with the actors and scenery projected into the theatre making a totally life-like effect. A little later will come the holographic video cassette, in full colour with stereo sound. Perhaps soon after the turn of the century you will be able to watch a film or a rock concert in vivid three-dimensional detail in the middle of your living room floor.

Aids to enjoyment
The digital watch combined with a calculator; the small screen portable TV/cassette player; the videogame; the personal stereo/radio; these are all remarkable electronic advances of the past few years. But they are only a taste of what is to come.

★ Microcomputers will shrink in size so that they, like the calculator, will fit alongside a wristwatch.

★ The pocket-sized television set will be on sale, in colour.

★ Combined with a microcomputer you will have a portable VDU/VDT and processor, though the inputs will not be the typewriter-style

▲
Twenty years ago, a calculator small enough to fit next to your watch would have been unheard of. In twenty years' time, even this will look clumsy compared to the possibility of a whole tiny computer on your wrist.

keyboard of today.

★ Sophisticated electronic games will emerge in which the player uses not manual or reaction skills as for space invaders but brainwaves to participate. These so-called biofeedback machines depend on self-control and a high degree of commitment.

★ And needless to say, television

sets will come in all shapes and sizes and incorporate high-resolution scanners with 1125 lines to make for crystal clear pictures.

Wall-sized screens, too, seem likely to emerge as a practical possibility for high resolution images long before the century is out.

◄ To get a 3-D picture of the model house a special photographic plate or 'hologram' is made using a laser beam. The plate just looks like an ordinary piece of glass until another laser beam is shone on to it. Then you can see the 'solid' house again – looking real enough to touch.

► A video cassette recorder that uses a 'silver disc' to play back recordings on TV. The disc is 'read' using a laser.

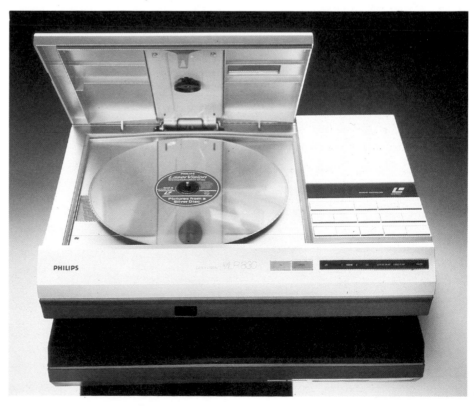

Leisure and education

Olympics 2000

Sport has already had its share of technology. Computers are programmed to work out athletes' training schedules and diets; medical researchers and doctors study the problems of injuries, performance under stress and surgery for correcting damaged tissues. There has also been considerable progress in designing better running tracks, tennis courts and sports fields, not to mention flexible fibre glass poles for pole vaulters and protective lightweight helmets for football players.

This trend of linking science and technology to sport is bound to continue. Experts predict that by the end of the century a combination of natural aptitude, strict training routines, totally supervised diets and lifestyle, and access to the best facilities will produce a mile world record of somewhere between 3.30 and 3.45 seconds and that the 100 metres men's freestyle swimming record will be 45 seconds. The high jump record will be in excess of 8 feet (2.5 metres), and the pole vault, using a graphite pole, over 20 feet (6 metres). There will be similar advances in the shot put, discus, javelin and hammer throws. The only event where the improvement will seem slender is the men's 100 metres. In 1921 the world record was 10.2 seconds. Sixty years later only three tenths of a second had been shed to 9.9 secs. It seems unlikely that more than another tenth can be shaved off this, though of course nowadays electronic timing means that records will be broken by hundredths of a second anyway.

There is one record that no amount of scientific training and no number of technological aids will improve by the end of the century. The men's long jump. The record of 29 feet 2½ inches (9 metres) is held by the American Bob Beaman who made his astonishing leap in the 1968 Olympics in Mexico City. At the time it knocked a fantastic 22

48

inches (56 centimetres) off the existing record and thus is reckoned to represent an improvement that you would only expect after eleven centuries of competitive performances.

Women's performances will possibly improve in an even more spectacular fashion than men's. Remember that today women are notching up running times as good as the best men in the 1970s. In those events where neither sex has a pronounced physical advantage the male-female performances are likely to be indistinguishable.

All the fun of the fair

When the fair comes to town today it still attracts young and old alike with its brightly painted collection of wood and steel helter-skelters, roller coasters, bumper cars and side shows. In future the trend will be towards high-technology game parks like Disneyland and Epcot in the US where you can go and be thrilled by amusements based on advanced engineering designs.

By the end of the century we can expect more fairs to follow the model of the World of Fun Amusement Park in Missouri. Here, for example, is the famous Orient Express roller coaster, a train of six cars which can reach a speed of nearly 60 miles per hour (100 kilometres per hour) as it swoops downwards and round a complete loop to be followed by another one almost immediately afterwards. Projects such as this require total safety in construction and control mechanisms combined with luxurious comfort and smoothness of ride. Many skills go into the design of these spectacular fairground structures: computers are needed to work out the thrusts needed to catapult the cars over the loops or to design the aerodynamically streamlined vehicles themselves. Inspection using X-rays is necessary to ensure that all the many welds in the superstructure are secure; the miles of wiring that are needed for these complex engineering systems have to be monitored constantly to avoid breakdowns, again using computers.

► Looping the loop in the funfair of the future. High technology is needed to design, build and control a complex roller-coaster system. Computers are used to maintain safety and tough materials specially designed to allow the cars' high-speed wheels to run with the minimum of friction.

CHAPTER 6
Space

The next two decades will see us getting to know very well indeed that tiny region of space that surrounds our own planet. We will learn a great deal about the practical problems involved in space engineering and research, and how human beings react to long-term trips in weightless conditions.

As we enter the 21st century, plans will form for building space platforms, even perhaps small colonies. But there will still be a drive ever outwards, to explore the distant reaches of our solar system, then to broader pastures, beyond our astronomical backdoor, to journey to other stars.

Space

Engineering in space
With the successful launch and recovery of the world's first reusable space vehicle, the Space Shuttle, a new era in spaceflight has emerged. Over the next two decades, the Shuttle family will grow, as will the range of activities it can handle. Many satellites will be launched from the Shuttle's loading bay. But those cargo doors will also be opening up to reveal other kinds of payloads: hardware and materials to be assembled into permanent structures. The Shuttle and its descendants will be the transport fleet of the future.

Here are some of the exciting projects that the 'space truck' as it has been called might bring to fruition.

Orbiting power plants
Plans already exist for building nuclear power plants in space – what is known as the shuttle-based reactor or SBR. The nuclear core, where the essential energy-producing atomic reactions take place, would be small enough to be carried in the cargo hold. Once in space it would be removed by special gantries and handling arms and held in position so that other parts of the reactor could be flown up in other shuttles and attached to it.

Once assembled, the reactor – a massive beehive structure – will remain in space to generate power. It would be near enough to Earth to be retrievable but far enough away to avoid risks of dangerous accidents and pollution.

Space platforms
Results from the Shuttle and from the Soviet linking of Cosmos 1267 to the smaller Salyut vehicle, suggest that, in the fairly near future, large space stations and space launching platforms will become a real possibility. The Shuttle removes the need to have massive thruster rockets to carry large cargoes, while the Soviet techniques for linking spacecraft to a central pod demonstrates the 'nuts and bolts' practicality of space stations.

Spacefaring passengers
NASA (the National Aeronautics and Space Administration) also has plans to use the Shuttle to carry people as well as scientific and engineering payloads. These will not of course be sightseers on an extra special excursion – at least not for some time to come – but people with jobs to do in making some of these high-technology projects come true. But they will not be trained astronauts either. It is thought that 10 000 people a year could be accommodated on these flights, at the rate of 74 per trip.

Space laboratories
To be in orbit in space is to be in a very special environment: outside is a near-perfect vacuum and zero gravity. Already the NASA Spacelab built for the European Space Agency is well on the way to using these facilities for a wide range of experiments. Some of these will be designed to find out more about purely scientific matters; others will be aimed towards applied research and development, perhaps with commercial applications.

Among the research projects already emerging is the investigation of the body's responses to weightless conditions. Shuttle scientists will

► Settlements in space. There have been many designs for space colonies, some huge structures that make our present technology seem almost childish. One such plan, a huge wheel-shaped island in space, rotating slowly one revolution per minute, would house 10,000 people in its rim. Here, with their artificial atmosphere and gravity, they would have shops and schools and factories; even agricultural areas to grow their own produce. Finding the raw materials to build such colossal structures will be a major problem.

be rocked to and fro in the 'space sled' to see how they react – an important study for future long-duration flights.

Within the various equipment racks on board space laboratories will be astronomy and physics experiments, chemistry and botany – in fact a whole range of intriguing projects. Among these will be the study of new types of manufacturing processes. It is thought that space conditions could lead to the development of revolutionary products: better fibre optic glass; improved semiconductors for use in computers; solar cells; even exotic space jewellery. Medical products too will be developed, especially new, purified drugs.

Towards the space settlement
As this century nears its close, scientists and technologists will be in a better position than they are now to contemplate the practical steps that need to be taken to provide a permanent setting for human beings in space. A place where future generations will be born and die, living, working, playing in a congenial environment they know as 'home'.

Space

◄Communications satellite in orbit.

War in space

It is inevitable that space technology will figure ever more prominently in the minds of military planners and strategists. Space has already become a useful 'territory' for peaceful purposes. But unfortunately, it also has a lot to offer in wartime. In the future it could well become a battleground, like a scene from *Star Wars* come true. Unless we can prevent such wars, it is unlikely we shall reach 2000 safely.

Modern defensive systems rely very heavily on communications. Space is invaluable in this respect. Both the US and the USSR have launched into orbit a large number of satellites designed to relay information around the world. This can be of different kinds. In the event of war or some other major international incident, the two superpowers want to be able to send information to and from the scene of the action, keeping governments, diplomats, troops and the population at large up to date with what is going on.

The deployment of military satellites has led to another trend which will also continue: the use of anti-satellite satellites (ASATs) sometimes called 'hunter killer' satellites. These are designed to detect enemy satellites, track them and target a missile to shoot them out of the sky. A variation on this which is also expected to play a part in future armouries is a small air-to-space missile carried under the wing of a fighter plane. The US has developed its F-15 ASAT to work in this way.

Electronic espionage

Satellites are a very effective way of gathering information such as the location of enemy forces and bases. And they orbit much higher than a spyplane, which is relatively vulnerable.

There will be increasing use of electronic intelligence gathering from space. Satellites will either orbit at fairly low altitude, providing high-resolution photographs of an area as they sweep over it. Or they may fly higher in geosynchronous orbit. At the moment, the pictures obtained from such orbits are inferior, but the coverage they provide is uninterrupted. In the future, the aim will be to improve image collection so that high-orbit satellites will give uninterrupted coverage, with a high degree of detail.

Zapping with lasers

The concentrated beam of light from a large laser can burn through almost any known substance. Alternatively, the laser light can be sent as fast pulses which can pierce a target like some futuristic hammer blow. Tests are being carried out on laser weapons that could either shoot down incoming enemy missiles, knock satellites from the sky or damage equipment and blind soldiers on the ground.

As these military lasers are developed, they will be deployed both from the ground and from the air. Ground-based weapons suffer the great disadvantage of having to pierce the Earth's atmosphere, which absorbs some of the very high energy needed to produce the beam. So research is being carried out with chemical lasers in aircraft and small, compact weapons to be housed in satellites themselves.

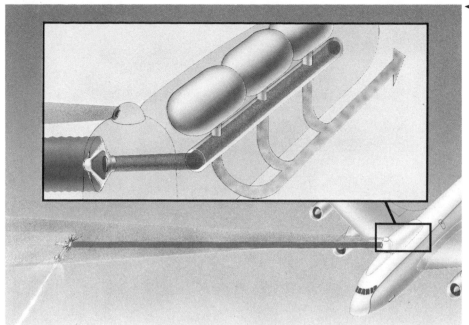

◄ Laser-beam weapon for short-range use.

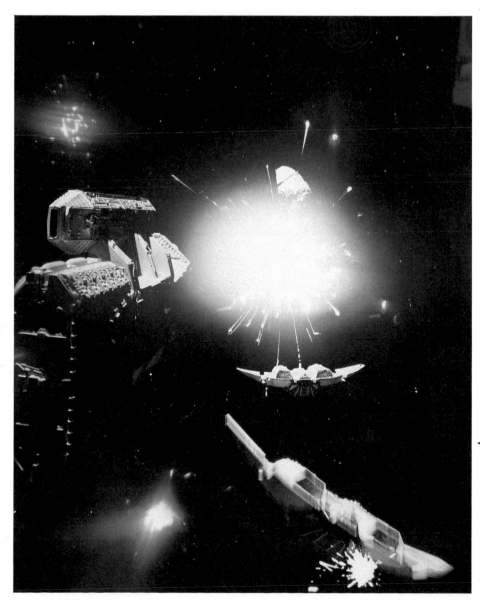

Particle beams

Looking beyond the laser, military planners are also thinking of the possibilities of using so-called particle beams as weapons. These are streams of electrons, protons or ionized particles (atoms that are stripped of electrons) charged up to enormous energies. Not only would these beams be highly destructive if they hit, say, a satellite or an inter-continental ballistic missile (ICBM), they would also be extremely fast in action. One prediction is that a particle-beam weapon could be directed in turn, at 100 ICBMs, each travelling at 5000 miles per hour (8000 kilometres per hour). When it reached the 100th one and destroyed it, this last missile would only have travelled 5 miles (8 kilometres) further than the first!

◄ A further refinement in laser 'death-ray' technology might be the laser battle station, sending beams to be guided by mirrors over thousands of miles. These would be 'locked-on', say, to an incoming intercontinental ballistic missile (ICBM) which would be destroyed literally in a flash.

Space

The outward urge

Space exploration is still in its infancy. Despite the manned moon landings, the unmanned Mariner probes to Mars, the Viking missions landing on the Martian surface and the Voyager trips to Jupiter, Saturn and the outer planets of the solar system, mankind has yet to go beyond the toddler stage of space travel. By the end of the century we will still, as space explorers, barely be able to walk properly – let alone run and jump – but we will have continued to direct our faltering steps outwards.

Long-distance space flight, that is beyond our solar system, towards other stars perhaps with their own planetary systems is still an unrealizable ambition. By the end of the century, though, we may see the firm preparations being made for a manned landing on our near neighbour Mars, and further unmanned probes being sent to the outer reaches of the solar system.

Developments in space hardware – rocketry, fuels, life-support systems – will continue apace. And schemes will emerge from the drawing boards for futuristic spacecraft using what are now unconventional means of propulsion. The fact is that the nearest stars are very distant from us. A spacecraft travelling at thousands of miles an hour would take tens of thousands of years to reach them. So manned long-distance flights will have to be in craft accelerated to speeds approaching the speed of light – 186 000 miles per second (300 000 kilometres per second)! An impossible task with present technology, but perhaps within the grasp of future generations using beams of photons or light particles emitted from their rockets like gigantic lasers.

While we are waiting for that future age to come, we may be lucky and make contact with civilizations living on distant star systems by means of unmanned vehicles. After Voyager leaves our solar system it will head outwards bearing a special recording, together with playing instructions, that provides a picture of life on Earth–its past and present. Some alien civilization may chance upon Voyager and find out all about us from this inanimate messenger.

Will the real ET please tune in

We can search for other intelligent life in the universe, if it exists, by means of radio. Indeed radio waves are a much faster method of trying to make contact than any form of space travel yet invented. On 16 November 1974 a radio signal was transmitted from the huge Arecibo Radio Observatory in Puerto Rico towards the star formation known as M13, which

lies 25 000 light years away (a light year being the distance travelled by light in one year – at a velocity of 186 000 miles or 300 000 kilometres a second).

The signal was very cleverly devised to contain a great deal of data within just 1679 bits of information. Should it be picked up by a distant civilization with a high technological capability they will learn about the basic chemistry of life on this planet; how genetic information is passed on by the DNA molecule; the world's population; roughly what we look like and what configuration our planets make around the Sun.

We do not know if this portrait of Earth has been received and/or understood. We could find out at anytime.

Life elsewhere: what are the chances?

By the year 2000 there may be a well-established SETI programme under way (Search for Extra-Terrestrial Intelligence). It will use some of today's large radio telescopes boosted by NASA's ultimate ambition, the multimillion dollar Cyclops project, an array of 1500 radio dishes each 108 yards (100 metres) across. How many alien ears might be out there in deep space listening to these signals?

Calculations are based on assuming that life elsewhere began on planets a bit like our own – which may or may not be true. Taking just our own galaxy – one of millions in the universe – there could, it is thought, be a million such planets in the Milky Way. So, by the same arithmetic, there are possibly billions of other inhabited worlds. If you assume, though, that life could exist in environments totally different from ours, then the number is truly vast. Who knows, the year 2000 may be the year we finally make contact with extra-terrestrial life from some distant planet!

Glossary

Access code Special instructions for obtaining information from a computer's memory.

Aerodynamics How aircraft, missiles, boats, etc. behave in motion through air, space and water.

Analogue Analogue computers are designed to mimic real-life situations on a small scale. They produce models or analogues. Most computers are *digital*.

Artificial intelligence Using computers programmed to behave like humans, with the capacity to learn to improve their performance.

Astronaut A space traveller; the word literally means 'a star sailor'.

Astronomy The study of the motions, positions and behaviour of bodies in space such as planets, stars and galaxies.

Atmosphere The gas surrounding the Earth or any other body in space. Around Earth the atmosphere is approximately 78% nitrogen, 21% oxygen, nearly 1% argon and small quantities of carbon dioxide, neon, helium, krypton and xenon.

Atom The smallest part of a chemical element which can take part in chemical reactions. It consists of a core – the nucleus – around which spin electrons.

Automation Using machines, including computers and robots, to reduce or replace the work of humans.

Battery A device for making and storing electricity, consisting of two metal electrodes in an acid solution, such as lead in sulphuric acid.

Binary The number system used in computers, based on just two digits, 0 and 1.

Bio- To do with living things. Thus biology is the science of life; biotechnology uses living organisms to make useful products; bioengineers construct devices such as artificial limbs to fit a human body.

Biofeedback A technique for trying to get the mind to control some of the body's functions.

Bit A 'binary digit', 0 or 1, the two numbers used in the digital computer.

Botany The study of plants (it is part of biology).

Byte A computer term for a group of bits, often shorter than a 'word', that are processed as a unit. The larger the computer the more bits can be handled in one byte.

Chemistry The science of how substances are composed and how they act on each other in reactions.

Chip An electronic circuit consisting of thousands of components etched on a small piece of silicon.

Chromosome A tiny, string-shaped package of genes, within the nucleus of the living cell.

Circuit A pathway linking electrical or electronic components and through which a current flows.

Code A way of expressing information by symbols, used in computer programming.

Component A part. It often refers to the various items that make up an electronic device.

Computer An electronic machine which takes in data (input), processes this in a series of logical steps, and supplies the results as information (output).

Conductor A material which allows electric current to pass through it, such as copper wire.

Cryosurgery The use of very low temperatures for surgical operations. The surgeon's scalpel or knife is an 'ice lance', a probe cooled to temperatures of $-50\,°C$ or below.

Data All the facts, figures and other information processed by a computer. It means literally 'things given'.

DBS Direct broadcasting by satellite. The use of satellites to relay TV broadcasts directly into the home.

Digital The commonest type of computer, using binary numbers to represent data.

DNA Deoxyribonucleic acid, the molecule shaped like a double screw (or helix) that carries genetic (inherited) information within the cell nucleus.

Drug A substance used to treat illness.

Electricity A form of energy based on the action of positive and negative charges. Electricity can be static or in motion as a current.

Electrode A piece of material that conducts electricity and is used to allow current to enter or leave a device such as a battery.

Electron A constituent of all *atoms*. Electrons orbit the central *nucleus*.

Electronics Using devices such as *semiconductors* to make equipment dependent upon the controlled movement of *electrons*: from computers to radios; stereos to watches.

Enzyme A substance produced by living cells which acts as a catalyst to promote biological reactions on which life depends.

Feedback An arrangement such as you find in industrial robots whereby actions are coupled to 'experience'.

Friction The force of resistance between two surfaces, such as a tyre on a road or a space vehicle moving through the atmosphere.

Galaxy A large collection of stars such as our own Milky Way. There are millions of galaxies in the universe, each containing around 100 000 million stars. The nearest galaxy to ours is over a million light years away.

Genes Structures of the molecule *DNA* situated within the *nucleus* of the living cell and controlling all its properties and behaviour.

Genetic engineering Sometimes called 'recombinant *DNA* technology'. The manipulation of genes to produce useful substances or to correct inherited medical disorders.

Geosynchronous Applied to the orbit of a satellite when it is circling the Earth in exact time with the Earth's own rotation. Thus it appears from Earth to be stationary ('geostationary').

Gravity The force that attracts bodies in space to each other, such as the Moon to the Earth or the Earth to the Sun.

Hardware The machinery of the computer as opposed to the program or *software*.

Helium A very light, non-inflammable gas, useful in balloons and airships.

Hologram Literally a 'whole image'. A picture in 3 dimensions – length, breadth and depth – produced by splitting a laser beam and throwing the image on to a photographic plate.

Hormones Common substances in the body which cause or regulate many bodily processes. For example, insulin produced in the pancreas controls the amount of sugar in the blood.

Hydrogen A light, colourless, tasteless, odourless gas. It is inflammable. It is also the most abundant element in the universe.

Impulse A short-acting force such as a rapid burst of electricity.

IT 'Information technology', a general-purpose phrase to cover the use of computers, TV, VDU/VDTs, telephones, VCRs, satellite, cables and other communications methods on a widespread scale. Whatever the machinery, the common ingredient is information, be this on tape, film or over a loudspeaker.

Keyboard An important device for putting information into any system such as a word processor or computer. It is an arrangement of finger-operated keys, the best known of which is named after the top left-hand row of letters on a standard typewriter: 'QWERTY'.

Language The specially devised code of a computer, written in binary numbers. Some well-known examples are BASIC and FORTRAN.

Laser A word made from the initials of light amplification by stimulated emission of radiation.

It is a very powerful, coherent beam of light.

Light year A measure of distance (not time) used by astronomers. It is how far light travels in one year, roughly equal to 10 million million kilometres.

'Maglev' Short for 'magnetically levitated': a method of keeping a vehicle such as a train raised off a rail by an arrangement of powerful magnets.

Mainframe The largest of computers.

Memory That part of a computer where information is stored ready for later retrieval.

Microcomputer (microprocessor, or 'micro') A computer built on a single chip. These are the smallest computers available.

Minicomputer The middle of the size range of computers between the mainframe and the micro.

Molecule An arrangement of *atoms*.

Monorail A single rail, such as one finds with *maglev* train designs.

Network/networking Linking together a number of computers which can intercommunicate.

Nuclear reaction By changing the state of the nuclei of *atoms* of certain substances it is possible to set off a series of reactions in a 'chain'. This releases vast amounts of energy which can be used for weapons of destruction or to provide electricity.

Nucleus The core of all *atoms*, consisting of neutrons and protons. The term can also refer to the centre of the living cell where genes are situated.

Orbit The path taken by one body round another, such as a satellite round the Earth.

Pancreas An organ of the body which lies across the back of the abdomen, and produces the hormone insulin.

Particle A basic unit of matter such as an *electron* or a *proton* which goes to make up the *atom*.

Payload The cargo carried by, say, a space vehicle. The Space Shuttle for example has had payloads of scientific experiments and satellites. Missiles carry explosive warheads as their payloads.

Photosetting A printing term, referring to the use of photographic film instead of metal type.

Physics The study of the properties and behaviour of matter and energy.

Pneumatic Operated by air under pressure – compressed.

Program The set of instructions which controls the operations of a computer. The person who writes it is the 'programmer'.

Proton One of the constituents of the *nucleus* of the *atom*.

Prototype A first or early example of something such as a test model of an engineering structure or a vehicle.

Resolution The extent to which an image can be made out by the eye.

Robot Originally a 'slave' or 'worker'. It has now come to cover a variety of machines: mechanical toys, automatic devices, industrial handling equipment and so on.

Robotics The study of robots.

Satellite. A body in *orbit* around another, such as a moon round a planet, or a communications vehicle round the Earth.

Semiconductor A material, such as silicon, which only partly conducts electricity. Its properties fall between those of a *conductor* and an insulator.

Sensor An instrument for detecting and measuring the properties and behaviour of things in its vicinity. Sensors can for example determine closeness, temperature or pressure and relay this information to a computer in an industrial robot.

Servomechanism An auxiliary or slave power mechanism which converts low-powered into high-powered motion.

Side effects The effects of a medicine or drug other than those of treating illness. Thus a normal dose of a pain killer such as aspirin can have undesirable side effects on the stomach by irritating it.

Silicon A chemical element which has semiconducting properties, used widely in making chips for electronic devices.

Solar system The Sun and its nine planets: Mercury, Venus, Earth, Mars, Jupiter, Saturn, Uranus, Neptune and Pluto.

Solar cell A device for turning the energy from the Sun into electricity. Sometimes called a photo ('light') voltaic cell.

Superconductor Material that conducts electricity superbly well, with practically no resistance. To achieve this, it has to be cooled to very low temperatures, in fact to levels near absolute zero – 273.15°C.

Synthetic Made up artificially, not naturally.

Tape A magnetically coated material for recording sound, sound and pictures or other kinds of information such as computer data.

Technology The application of scientific principles to practical problems and projects.

Tele- Means 'at a distance', hence telephone (sound at a distance) and television (seeing at a distance). Many words have been made with this in recent years, including teledata (using TV for newspaper-style information) and teleconferencing (long-distance meetings without needing to travel). The catch-all term for them is telecommunications.

Transistor A semiconducting device for switching or amplifying signals. Used in radios and earlier generations of computers.

Vacuum Space in which there are no, or at least very few, molecules of any elements, such as one finds in outer space.

VCR Video cassette recorder.

VDU/VDT Visual display unit/terminal. The most common sort is the TV screen which may be linked to a computer or other information source to provide a display of data, as text, or diagrams. Word processors have purpose-built VDU/VDTs.

Word processor Not just a typewriter but a whole system for preparing, editing, storing and printing text. It consists of a VDU/VDT, a typewriter-style keyboard, memory and printer. WPs will be an important feature of the office of the future.

Index

Page numbers in *italics* refer to relevant illustrations and captions.

Picture acknowledgements

Atari 10 bottom right **BBC** 10 left inset
British Telecom 25 right inset **Daily
Telegraph Colour Library** 2 top, 4 top
right, 7, 28, 29, 49 **Mary Evans Picture
Library** 13 top, 45 top **Hugin Cash
Registers Limited** 37 top right **Mullard
Limited** 13 bottom right **Multimedia
Publications** 6 top left, 17, 34-35 **Philips
Laser Vision** 47 bottom **Rex Features** 4 top
left, 6 centre, 6-7, 39 top, 39 bottom, 43, 44-45,
46, 58 centre **Ann Ronan Picture Library**
12 **STC** 11 bottom left **Science Photo
Library** 4 centre, 4 bottom, 13 bottom left, 16
top left, 24-25, 31 left, 31 right, 36, 38, 40, 58
bottom **Space Frontiers** 1, 54 **Spectrum

Colour Library** 9 left, 10 top right, 55
Frank Spooner Pictures 24 inset, 37
bottom left **Tony Stone Photolibrary,
London** 11 bottom right, 26-27, 37 bottom
right, 37 top left **Texas Instruments** 9 top,
42 left **Vision International** 9 right, 16
bottom, 33 bottom, 45 bottom **Young Artists**
52-53 **ZEFA** 3, 6 bottom, 10 top left, 11 top,
25 centre inset, 33 top

Artwork by **Mulkern Rutherford** and **John
Strange**

Multimedia Publications (UK) Limited have
endeavoured to observe the legal
requirements with regard to the suppliers of
photographic and illustrative materials.